Light

A Spectrum of Opportunity

Library of Congress Control Number: 2023049314

Published by SPIE
P.O. Box 10
Bellingham, Washington 98227-0010 USA
Phone: +1 360.676.3290
Fax: +1 360.647.1445
Email: books@spie.org
Web: http://spie.org

Printed and bound in the United States of America.

Light
A Spectrum of Opportunity

SPIE PRESS
Bellingham, Washington USA

Authors

Written by:

Agata Azzolin

Isinsu Baylam Toker

Andrea Curatolo

Danuta Sampson

Gavrielle Untracht

Philip Wijesinghe

Cover designed by:

Marta Jakubowska

"Did you know" cards designed by:

Natália Nagy

Coordinator:

Danuta Sampson

Find the "Did you know" cards online at this link:

https:/spie.org/Samples/Pressbook_Supplemental/CB03_sup.zip

Project supported by a 2022 SPIE Outreach Grant

Introduction

The book you have in your hands will help you to love optics and photonics—the science and application of light. It starts with short sentences and beautiful pictures that demonstrate the presence of light in our lives. Later, it moves to more complex descriptions. The last pages summarize the basics of optics and photonics and job opportunities in these exciting fields. You will also find a family card game, beautifully illustrated by Natália, that will help you further enhance your knowledge about the topic.

The intention of the book is to move from simple to complex information and make the book a longstanding friend for everyone. For example, a child may only like to look at pictures and move to more advanced descriptions later in life. That is absolutely fine. Life is a journey, and we need different stimulations at different stages of life to learn and remember.

If, after reading this book, you would like to explore more optics and photonics through hands-on activities, the book *Discovering Light: Fun Experiments with Optics*, edited by Maria Viñas Peña, is a beautiful source of inspiration for experiments.

Acknowledgments

To Eugene Arthurs, Kishan Dholakia, Halina Rubinsztein-Dunlop, David Sampson, Anna Szkulmowska, and Maria Viñas Peña for feedback about the book you are holding; Jiawen Li, Sapphire Lally, and Szymon Tamborski for help with the "Did You Know" game linked at the front of the book; and to everyone above for championing optics and photonics worldwide.

To SPIE and Optica for inspiration, opportunities, and resources to deliver optics and photonics outreach at schools worldwide.

To our colleagues and friends who have been with us on the journey to promote the beautiful world of optics and photonics—too many to be listed here, but you know who you are, don't you?!

To students in schools worldwide, whose eyes lit up when we did optics experiments together.

To you, the reader, we hope you will enjoy this journey and become inspired.

We dedicate this book

To everyone, but in particular, to Jesse and the other children. We hope you will become inspired by light, its beauty, and its endless presence in our lives.

What is light?

Light is everywhere.

Light is how we start the day.
The sun rises and shines
light on the beautiful world
around us each morning.

Light is how we end the day.

Each evening the sun sets and gives us a beautiful light show.

Light is colorful.

When light meets the water droplets in the air, we see a rainbow.

Sometimes light behaves like a small object that we call a photon.

Photons bounce off other objects like tiny bouncy balls.

Sometimes light behaves like a wave.

When waves encounter each other, they superimpose on one another like ripples in water.

Light travels in straight lines.

Light can bounce off objects.

When light bounces off an object, like the surface of a lake, we call it reflection.

Light can travel through some solid objects, like windows. If light can travel through an object, we say the object is transparent.

Look at the spoon in the glass. Do you see that it doesn't look straight? This is an optical phenomenon called refraction.

Refraction happens because light bends where transparent objects, like glass and water, meet.

Light is energy.

Plants use energy from sunlight in a process called photosynthesis. Photosynthesis is a plant's way of making food for itself.

Some buildings use solar panels to harvest energy directly from sunlight. Solar means "from the sun." These panels catch sunlight and convert it into the electricity we use to power our homes.

Our bodies use sunlight too. The rising and setting of the sun tells our bodies when to eat, sleep, and wake up.

Sunlight helps our bodies produce vitamin D, which keeps our bones strong and makes us happy.

Humans have been able to create and control light for thousands of years.

We call it artificial light.

FIRE

As old as humankind.

OIL LAMP

Introduced in 10th millennium BCE during the stone age.

CANDLE

Introduced in the fourth century BCE in ancient Rome.

LIGHT BULB

First invented in the 19th century in the United Kingdom.

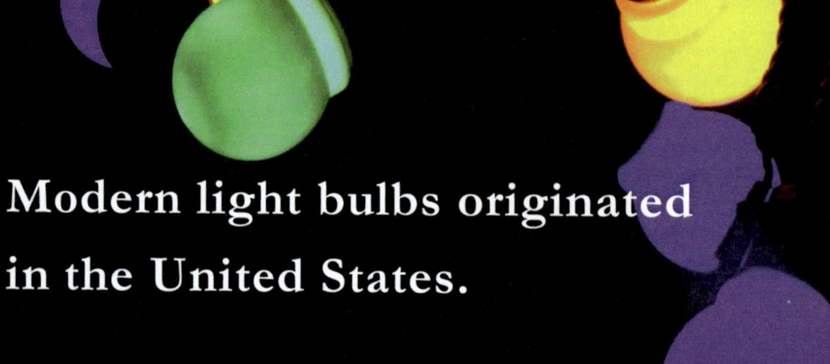

Modern light bulbs originated in the United States.

LASER

Invented in the 20th century in the United States.

Light bulbs can produce different kinds of light.

When you want to play games at night, you need a bright light.

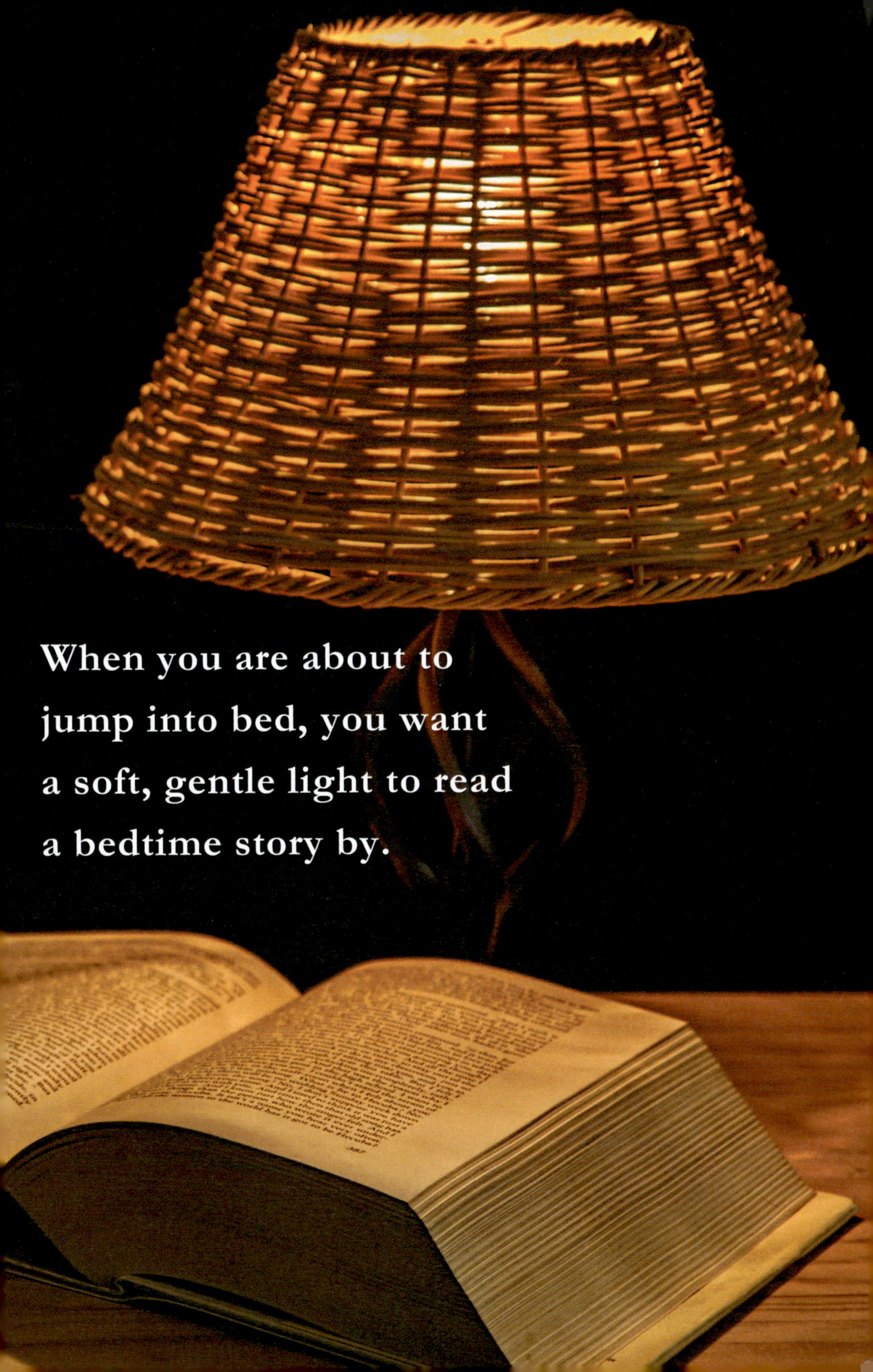

When you are about to jump into bed, you want a soft, gentle light to read a bedtime story by.

Car and bike headlights shine brightly at night, making us more visible in the dark.

Light in a microscope helps us see things that are very small, like cells.

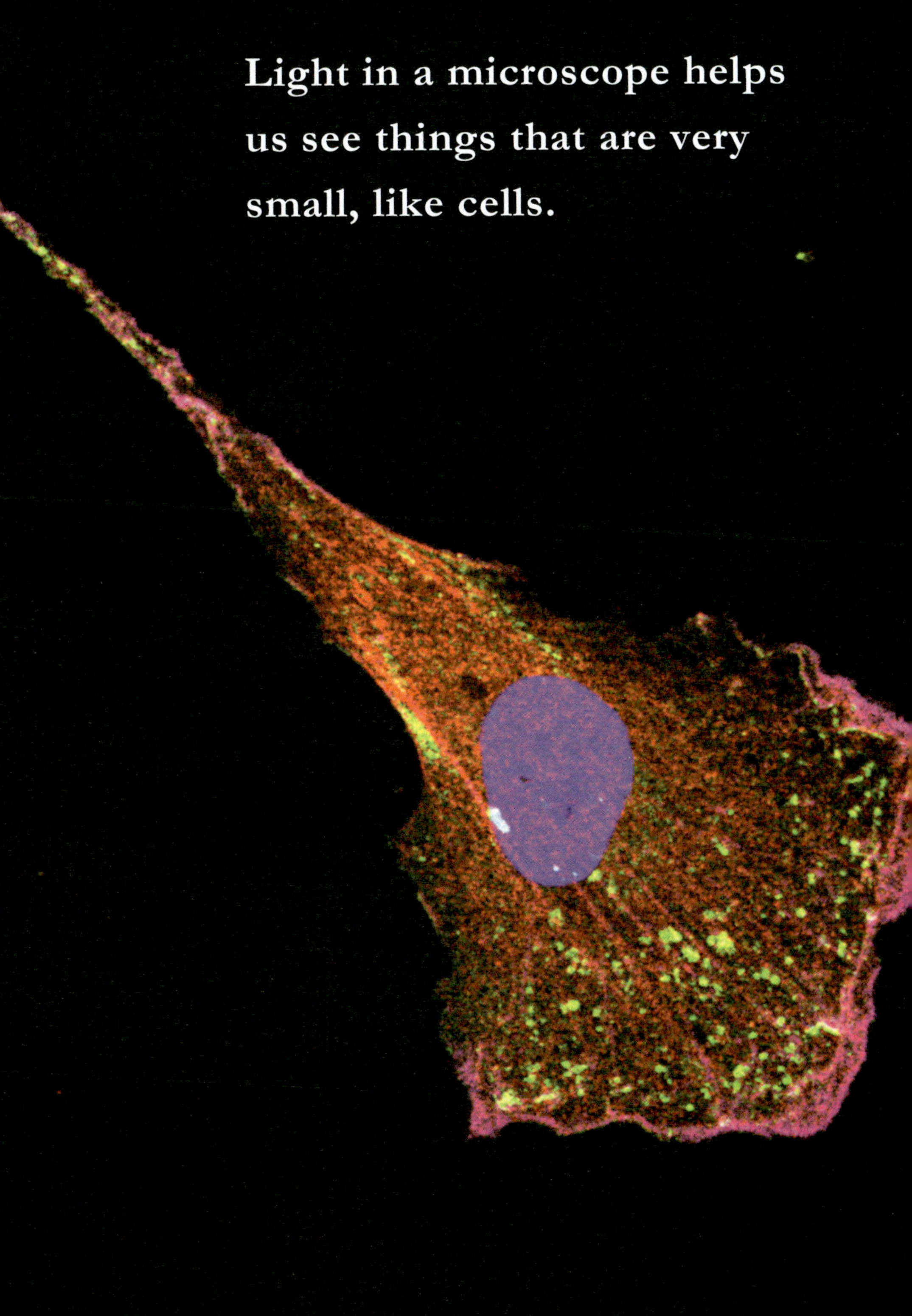

Light in a telescope helps us see things very far away, like stars and galaxies.

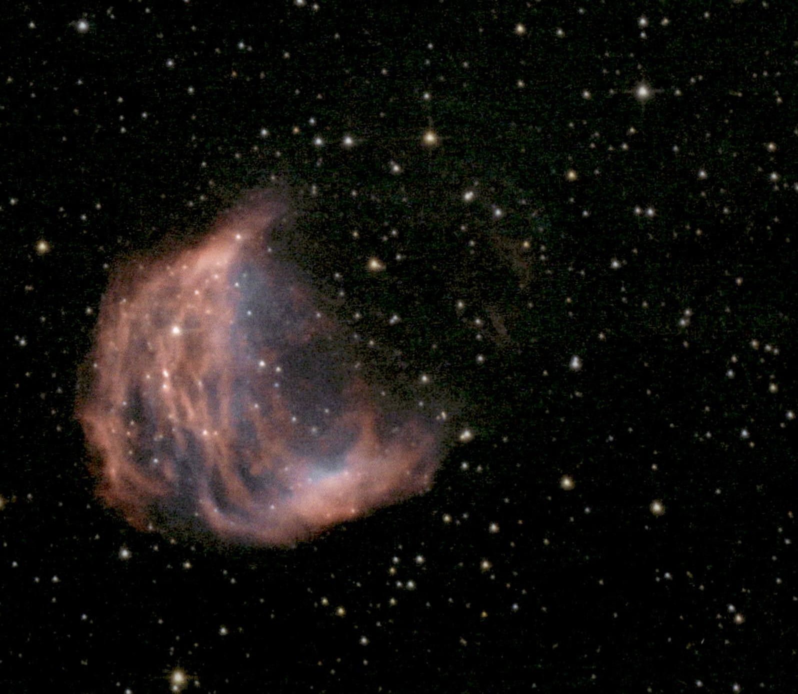

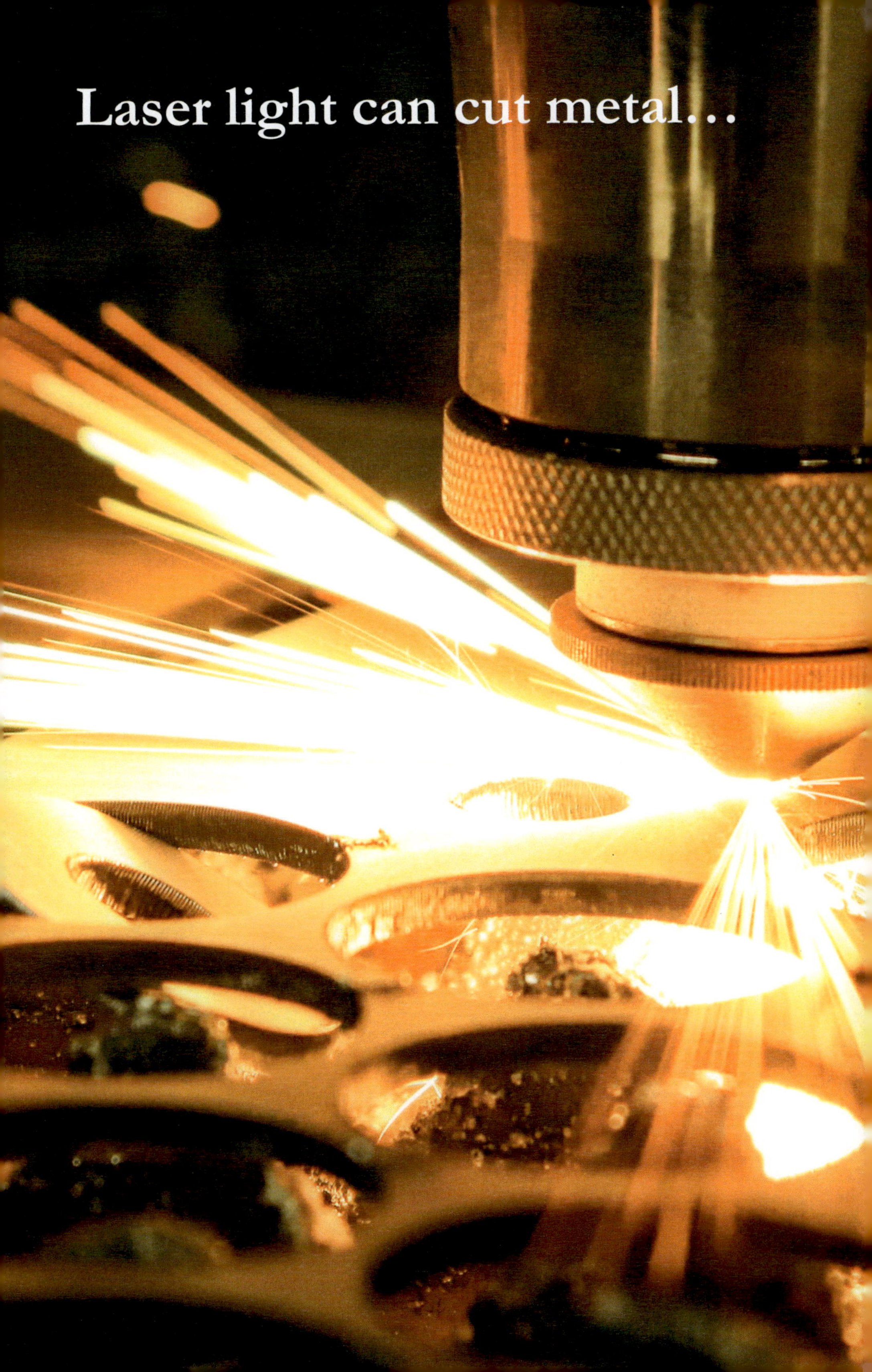
Laser light can cut metal...

... to shapes that can be used, for example, to make a car.

Light-based cameras and sensors on planes and drones can tell farmers which crops need to be watered...

... and then the crops can produce food.

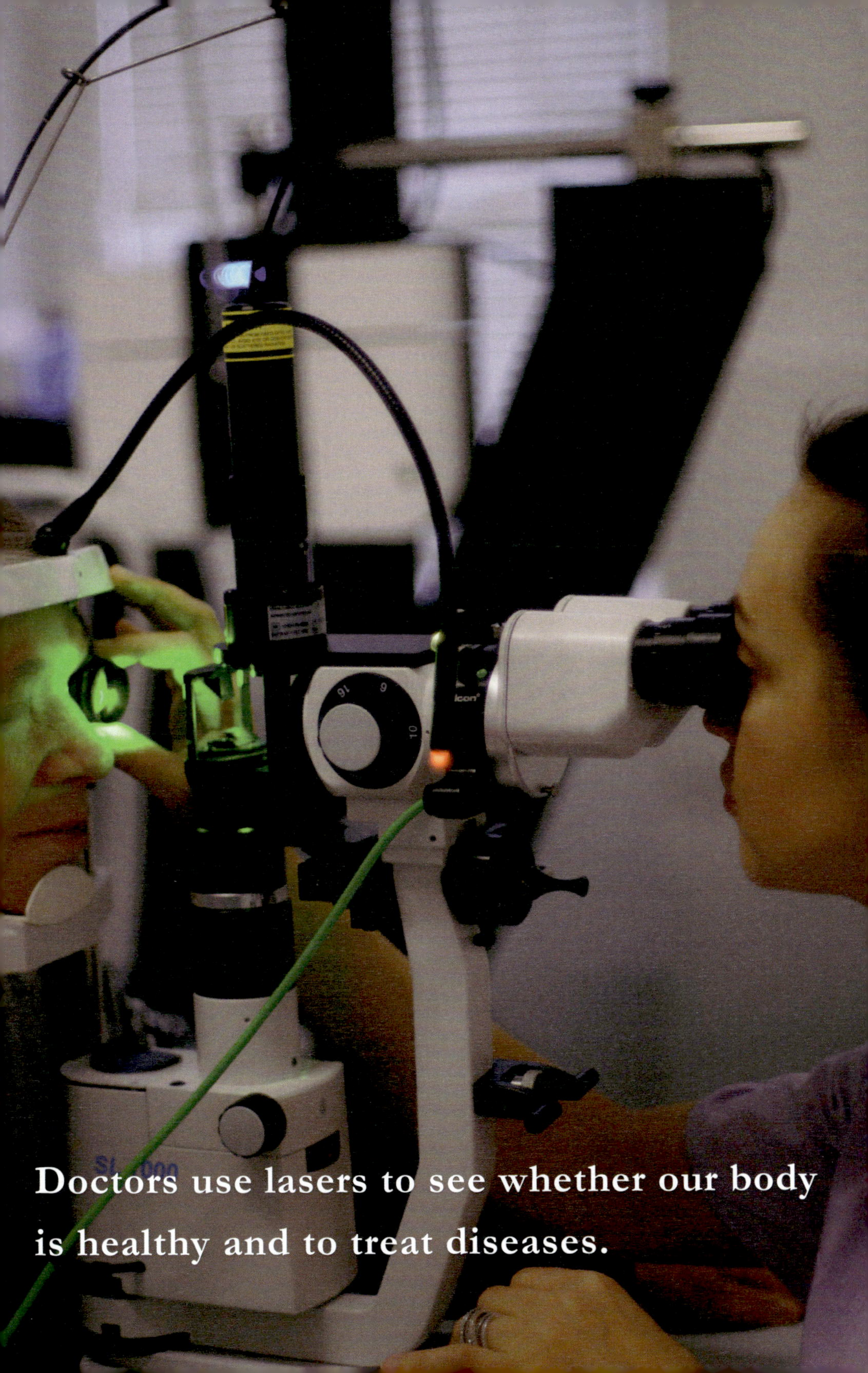

Doctors use lasers to see whether our body is healthy and to treat diseases.

The Internet works over long distances by using light traveling through glass wires called optical fibers. This helps us stay connected with friends.

Light is contributing to the quantum revolution
toward new technology,

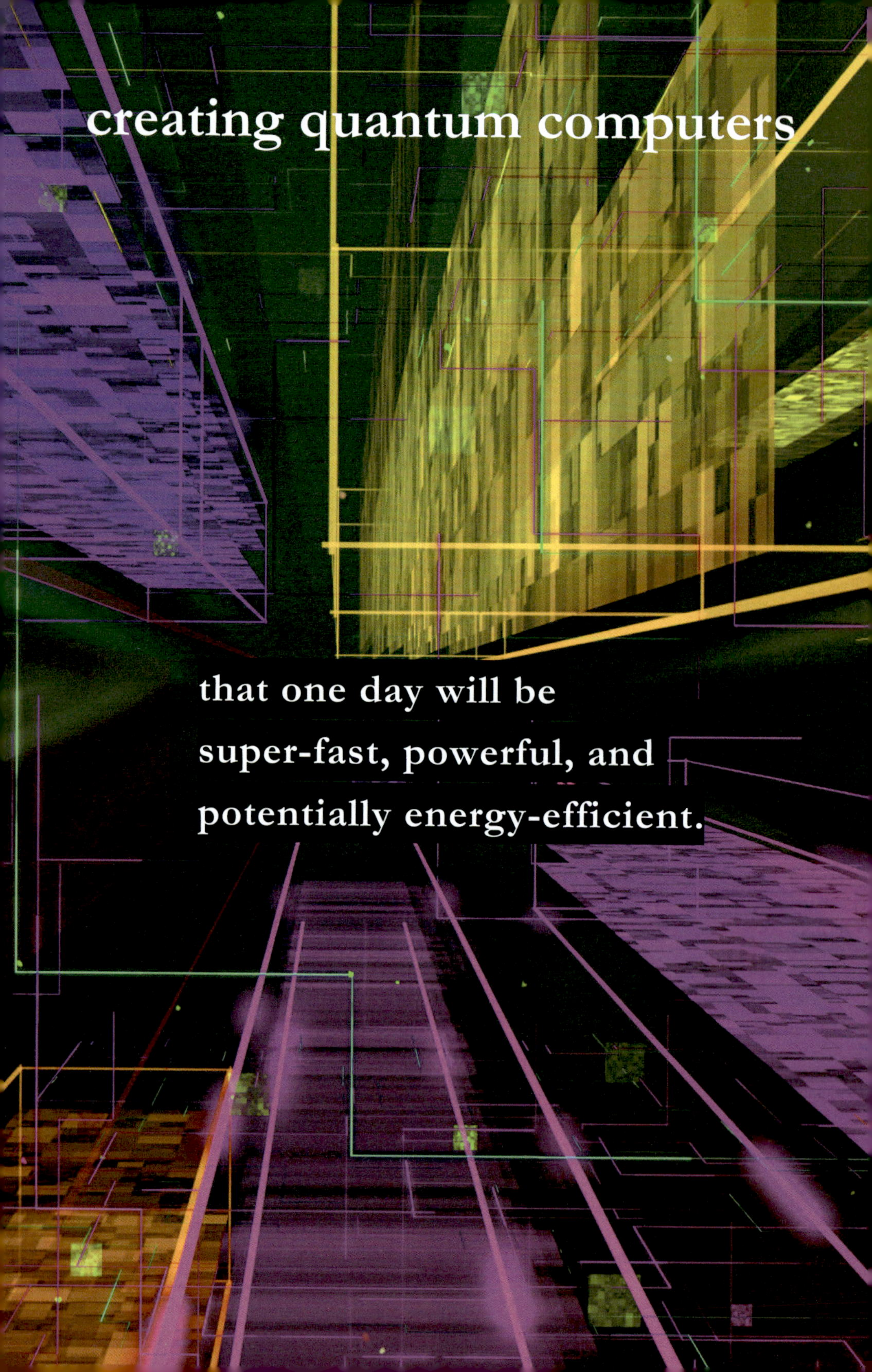
creating quantum computers
that one day will be
super-fast, powerful, and
potentially energy-efficient.

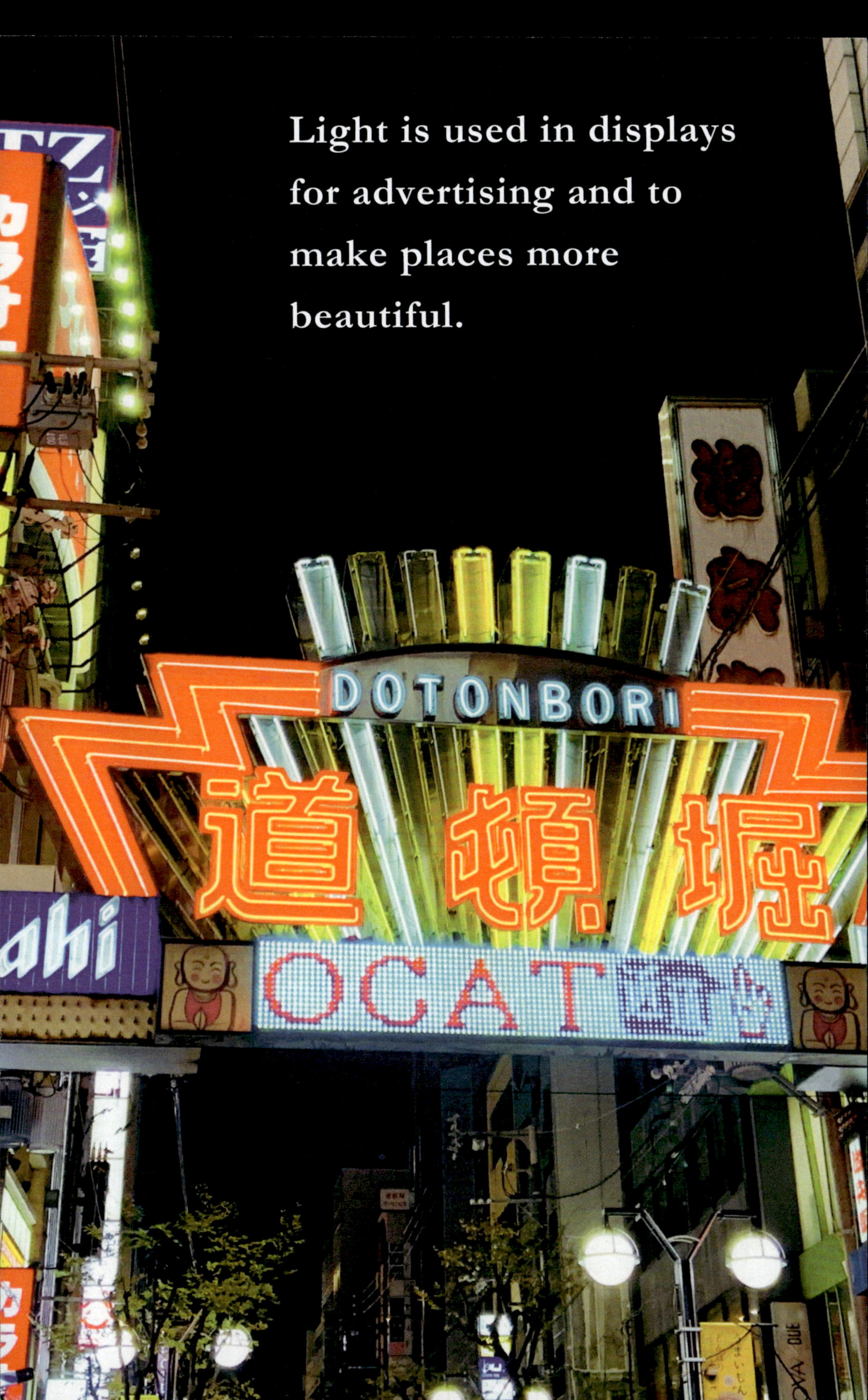

Light is used in displays for advertising and to make places more beautiful.

We can create art with light.

Light inspires us,

creating opportunities to see beautiful optical phenomena, including the northern lights.

Without light, you could not read this book.

Light is a form of **energy**.

The Sun is a very important source of light energy because there would be no plants, animals, or humans without it.

But humans can also make light. We call it **artificial light**. The most common **artificial light** sources are fire, light bulbs, flashlights, and lasers.

Light is made up of tiny particles called **photons**.

Light—a more advanced summary

But light is also a **wave**—an electromagnetic wave.

We need both descriptions to understand light.

Light travels in a straight line at a speed of **299,792 kilometers per second** (in vacuum). You could go around Earth eight times in a second at this speed!

An important property of a light wave is its **frequency** or **wavelength**.

The **wavelength** is the distance between two crests of the wave. The **frequency** is how many crests pass a specific point in 1 second.

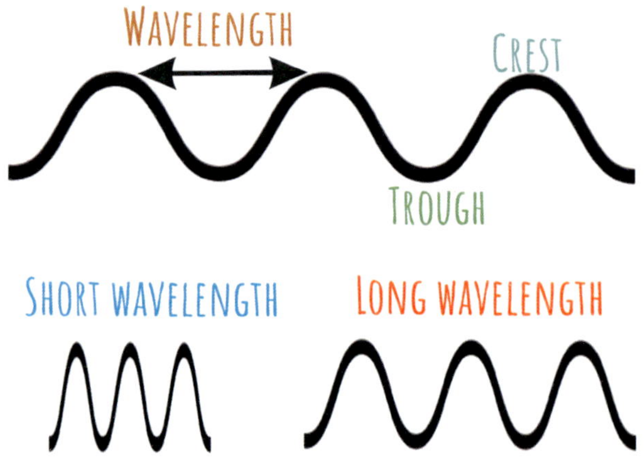

A light photon can be described by its energy, which depends on its frequency and wavelength. A higher light frequency (or shorter wavelength) means a higher energy.

The **color** of the light depends on its wavelength.

The range of wavelengths that we can see is called the **visible spectrum**. For the light we can see, violet light has the shortest wavelength (400 nanometers) and highest energy. Red light can have a wavelength of up to 700 nanometers—the lowest energy.

One hundred nanometers is more than 1000 times smaller than the width of a human hair!

Although we cannot see it, we know there is light outside the visible spectrum. **Ultraviolet light** (an even shorter wavelength than visible violet) is responsible for skin tanning and sunburn on a sunny day. **Infrared light** (a longer wavelength than visible red) can warm us, and we have invented special instruments to use infrared to see better in the dark.

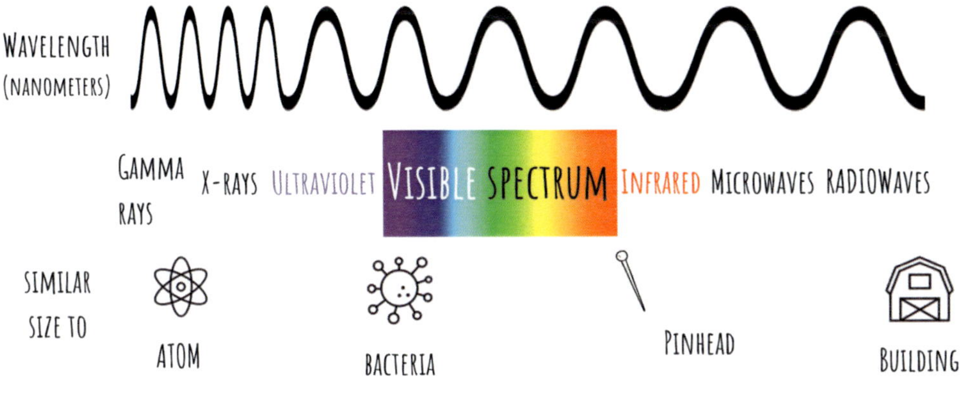

People have studied light for centuries, but only in the last 60 years have optics and photonics become prominent fields. Light, optics, and photonics impact almost every aspect of our lives, from entertainment to medicine, communications, energy, agriculture, art, and culture.

Optics is a general area of physics covering a wide range of topics related to studying light. Optics includes subfields such as geometrical, physical, fiber, quantum, and biomedical optics.

Optics and Photonics

Photonics deals with the science behind and applications of the generation, detection, and manipulation of light. The term "photonics" came into wider use in the 1960s with the invention of the laser and later the semiconductor laser. It was originally intended to describe a field where the goal was to use light to perform functions traditionally accomplished using electronics, thus the name.

Optics and photonics play important roles in how we stay safe, communicate, travel, diagnose and treat diseases, monitor the environment, and generate clean energy and art, among other things. The presence of optics and photonics in our lives is **HUGE**.

We always need more people to work in the field.

Careers in Optics & Photonics

With expert knowledge in optics and photonics and a love for light, you could work in:

Communications: telecom systems, cell phones, displays, fiber optics

Medicine: diagnostics and imaging, therapies, gene modification, drug development

Sensing: satellite and astronomical imaging, security systems, cameras

Manufacturing: process control, laser machining, semiconductor fabrication

Energy generation: solar panels, energy delivery by lasers, laser fussion

Security: guidance and control, displays, weapons

Research: university faculty, industrial and government scientist

**Light,
Optics, and
Photonics
are the future.**

PHOTO CREDITS in order of appearance: **Sunrise** (Mark Lucey), **Sunset** (Danuta Sampson), **Rainbow** (Chris Stockbridge), **Light is a particle** (László Nagy), **Light is a wave** (Hilmi Isilak, Pexels/Creative Commons CC0), **Light traveling straight lines** (Danuta Sampson), **Light reflection** (Isinsu Baylam Toker), **Light traveling through the window** (Danuta Sampson), **Light refraction** (Danuta Sampson), **Light is energy** (Marcin Grzybowski), **Photosynthesis** (Danuta Sampson), **Solar panels** (Danuta Sampson), **Clock** (Agata Azzolin), **Happiness** (Danuta Sampson), **Campfire** (Karol Karnowski), **Oil lamps** (Sandra Petersen, Pexels/Creative Commons CC0), **Candles** (Isinsu Baylam Toker), **Light bulb** (Wiesława Bukowska), **Laser** (Jovydas Pinkevicius, Pexels/Creative Commons CC0), **Bright light** (Daniel Reche, Pixabay/Creative Commons CC0), **Gentle light** (Steve Buissinne, Pixabay/Creative Commons CC0), **Lights on cars** (Karol Karnowski), **Microscope image of retinal pigment epithelial cells** (Berna Morava), **A spiral galaxy NGC2395** (Maciej Jarmoc), **Laser cut metal** (Dmitrii Bardadim, Pixabay/Creative Commons CC0), **Car** (Avinash Patel, Pexels/Creative Commons CC0), **Soil monitoring drone** (DJI-Agras, Pixabay/Creative Commons CC0), **Fields** (Agnieszka Bukowska), **Eye care** (Paul Diaconu, Pixabay/Creative Commons CC0), **Optical fiber/Internet** (Danuta Sampson), **Quantum revolution** (Pete Linforth Pixabay/Creative Commons CC0), **Quantum computer** (Oleg Gamulinskiy Pixabay/Creative Commons CC0), **Light in displays** (Isinsu Baylam Toker), **Light in art** (Isinsu Baylam Toker), **Northern Lights** (Chris Stockbridge)